Beginners Quick and Easy Guide to Making Money in

AGRICULTURE

ALEX NKENCHOR UWAJEH

LEGAL DISCLAIMERS

All contents copyright © 2017 by **Alex Nkenchor Uwajeh**. All rights reserved. No part of this document or accompanying files may be reproduced or transmitted in any form, electronic or otherwise, by any means without the prior written permission of the publisher.

This book is presented to you for informational purposes only and is not a substitution for any professional advice. The contents herein are based on the views and opinions of the author and all associated contributors.

While every effort has been made by the author and all associated contributors to present accurate and up to date information within this document, it is apparent technologies rapidly change. Therefore, the author and all associated contributors reserve the right to update the contents and information provided herein as these changes progress. The author and/or all associated contributors take no responsibility for any errors or omissions if such discrepancies exist within this document.

The author and all other contributors accept no responsibility for any consequential actions taken, whether monetary, legal, or otherwise, by any and all readers of the materials provided. It is the reader's sole responsibility to seek professional advice before taking any action on their part.

Reader's results will vary based on their skill level and individual perception of the contents herein, and thus no guarantees, monetarily or otherwise, can be made accurately, therefore, no guarantees are made.

BEGINNERS QUICK AND EASY GUIDE TO MAKING MONEY IN AGRICULTURE

CONTENTS

INTRODUCTION 7
   Agriculture and Huge Growth Potentials 10
   Organic Produce 12
   Ethnic Diversity 13
   Artisanal Foods 14
   Special Dietary Needs 15

FOOD PROCESSING AND STORAGE FACILITIES 16
   Build a Recession Proof Business 19
   Understand Your Business's Components 20
   Planning and Management 23
   Insurance 26
   Multiple Markets from One Product 27
   Multiple Points of Sale 29

DIVERSIFY AND EARN HUGE INCOME 32
   Profiting from Under-Utilized Areas 33
   Sub-leasing or Share Farming 33
   Agistment 34

BEGINNERS QUICK AND EASY GUIDE TO MAKING MONEY IN AGRICULTURE

Profiting from Small-Scale Dual Farming 35
Profiting from Companion Planting........38
   Lavender..39
   Garlic 40
   Gourmet Mushrooms............................42
   Chamomile 43
Profiting from Herbs 44
Profiting from By-Products 45
   Compost 46
   Fertilizer..47
   Bees 47
   Worms 48

BUILDING A MILLION DOLLAR AGRIBUSINESS..49
   CASE STUDY #1: Environmentally-Friendly Permaculture Farming Success .50
   CASE STUDY #2: Cherry Grove Organic Farm..52
   CASE STUDY #3: Dairy Farming and Artisanal Cheese Making.......................53
   CASE STUDY #4: Crawfish and Rice Farming Success......................................55

CONCLUSION 57

INTRODUCTION

If you're like most people, it's likely you'll associate farms as being little more than money pits. There is a commonly held image of the type-cast hard-working farmers as they toil away at the land all year for a meagre profit and retire to a dilapidated farm house for a bit of rest, only to repeat the process year after year.

Movies and TV series about struggling farmers have done much to promote the typecast image of poor farmers, ranchers, and other rural people living on the land. As a result, the negative images of people trying to grow rich from agriculture seem to thrive.

However, what most people don't realize is that many agricultural businesses are now capable of becoming more profitable than they have been at any time in the past. The key to breaking the old stereotypical images is to take advantage of emerging trends in the market and learn to structure agricultural enterprises to protect them against difficult financial times.

In recent years there has been a significant increase in the demand for fresh, locally sourced produce right across the country. Restaurants, cafés and eateries focus strongly on creating menus that feature local ingredients. Grocery stores are keen to supply their customers with an expanding selection of fresh, locally-sourced produce.

Individual customers still flock to local farmer's markets to buy the freshest produce available direct from the grower. The internet has also made it significantly easier for farmers to promote awareness of their produce and announce availability of certain products throughout each season.

Those same individual customers are also driving an increase in demand for specialty products created using those same fresh ingredients, with many farmers reporting sales of homegrown and homemade preserves, pickles, jams, jellies, sauces, spices and other specialty products increasing steadily.

Thanks to improvements in agricultural technology and machinery, it's become easier to maintain and manage a farm. Tools and technology can improve crop yields, increasing the profits per square foot of growing space.

The advantages offered by modern technology combined with the willingness to forge an income from the land provide any agricultural business owner with the seeds needed to generate profits and create success.

Agriculture and Huge Growth Potentials

When the majority of people think about agriculture, they picture big farms spanning across large acreage. They imagine thousands of heads of cattle roaming across the landscape or they imagine fields filled with brown-gold heads of wheat waiting for massively expensive farm machinery to arrive for harvest.

What you may not realize is that it's often possible to make more money on a small farm with just an acre or two than on a massive farm with huge acreage.

Yet many people regard the idea of profitable micro-farming with skepticism or even outright scorn. They seem to believe the production can't possibly be high enough with only one or two acres to generate the revenue needed to remain profitable.

After all, commercial farmers take advantage of economies of scale and use something called commoditization to their benefit. This simply means that goods from a variety of different farmers or producers are pooled together and sold as commodities. Commoditization offers a larger commercial producer less variance in pricing from the buyer's perspective.

Small-scale farmers don't have the same economies of scale in their favor. Instead, profitable small-scale farming is all about working smarter – not harder.

The growth potential of any agricultural business is now more promising than ever before, thanks largely to the increasing demand for a broader range of fresh produce than ever before.

While many larger commercial producers may stick to growing commodity crops, such as wheat or soy or rice, a micro-farming operation has the flexibility and freedom to grow a range of income-producing gourmet or specialty crops that help to supply local demand.

Organic Produce

Consumers are becoming increasingly health-conscious and are more informed now about the food they choose to eat. The demand for organic produce, farmed without chemical fertilizers, pesticides or herbicides continues to increase steadily.

While most grocery stores readily offer organic produce to customers, many find it challenging to access locally certified organic producers and suppliers that can provide a reliable level of supply allowing them to keep up with demand.

Selling organically-grown fruit and vegetables from a roadside stand or direct at farmer's markets is an effective way to sell your produce. Direct selling also allows you the opportunity to set your own prices based on quality of the produce you're selling. You're not limited by the commodity prices set by commercial growers and sold in supermarkets.

Ethnic Diversity

Increasing ethnic diversity is partially responsible for fueling demand for specialty foods. The increase in popularity for Mexican, Latin American, Caribbean, Italian, Spanish, French, Greek, Korean, Chinese, Thai, Kosher, and Halal foods means grocery stores and restaurant owners must find reliable sources of local produce or resort to importing the ingredients they need to keep up with demand.

Where many specialty ingredients may have been imported in the past, savvy farmers are grasping the opportunity to grow many specialty items locally to supply local markets. The result is that consumers are prepared to pay slightly higher prices for locally grown fresh produce in preference over importing tinned or frozen versions of the same item.

Artisanal Foods

The past decade has also seen an enormous increase in demand for more artisanal and gourmet foods. Cooked and cured meats, dried fruits, cheeses, oils, vinegars, spreads, chocolates, confections, and other types of artisanal products provide new eating experiences. The ready availability of so many exotic fruits and vegetables further enhances the opportunity for consumers to experience and enjoy exotic or unusual flavor combinations.

Special Dietary Needs

There is also the increased awareness of foods required to serve those with special dietary needs. It's estimated that the food allergy and intolerance market in North America alone should exceed $24.8 billion by 2020.

The rise in the number of people with celiac disease is a primary driver for the growth in demand for gluten-free food options. People living with lactose intolerance also demand alternatives that allow them to eat the foods they enjoy without the health consequences.

FOOD PROCESSING AND STORAGE FACILITIES

Growing crops or raising animals on your farm successfully is just the beginning. Your goods also need to be packaged and shipped to their final destination.

The roles of agricultural processing, preservation and storage are increasingly important in any farm operation. It's not possible for any agriculture system to develop to any level of scale without having the capacity to process and store goods produced.

Food processing begins with field-edge processing and storage and continues through a number of stages until the final product is packaged and ready to be sold to the end consumer. Some common food processing storage facilities might include:

- Grain bins and silos
- Animal housing
- Fertilizer storage
- Food processing and packaging facilities

Investing in food processing and storage facilities offers farmers the opportunity to process much of their own produce directly. The actual final product you grow or raise on your farm will dictate the type of processing and storage facilities you may require.

Food safety should be any farmer's number one priority. A well designed food processing and packaging facility can help prevent product contamination. Raw goods receiving and storage sections within the processing facility should be separated from other areas to further reduce the risk of contamination.

It's also important to implement proper food handling and hygiene practices.

Food packaging is also an important aspect of any agricultural business. The packaging in which your final product is presented needs to provide important information to your customer.

If you're selling fresh fruit or vegetables directly to customers at a farmer's market stall, packaging may be as simple as a crate or display box placed on a table. However, if you're selling packaged goods intended for sale in a grocery store or other retail outlet, your packaging will need to include correct labeling that meets regulations.

Some typical information that may need to be included on food packaging labels include:

Product name

Quantity packed (by volume or weight)

Name and address of producer

Storage instructions

Always check the regulations in your own state regarding packaging and labeling regulations for your specific goods.

Build a Recession Proof Business

Hard work and willpower on their own won't create a successful agribusiness. The key to building a recession-proof business is to bolster revenue streams, plan for the worst, and mitigate any foreseeable risks that may arise.

While every farmer is at the mercy of unforeseen weather conditions, there are still plenty of things that can be done to reduce the risk of exposing your entire business model to a temporary decline in business.

Understand Your Business's Components

The key to building a successful business in any industry lies in understanding each individual component of the business's structure. It's easy to think of farming in simple terms of having to do little more than produce livestock or grow crops and find paying customers. However, there are more elements to take into account if you hope to be successful. These include:

- Business structure (sole proprietorship, partnership, LLC, corporation, or co-operative)
- Production of your final products
- Machinery, equipment and technology
- Packaging

- Finances, bookkeeping and accounting
- Sales and marketing
- Land and property
- Shipping, transport and logistics

Each individual component of any agricultural business has a critical role to play in your overall success. Take the time to work out exactly what crops you'll grow or what animals you'll raise.

The actual crops or animals you choose to farm will be heavily dependent on the size of your land, the climate you're in, and the soil conditions.

Make the decision about whether you'll buy land and potentially pay off a mortgage or whether you'll rent land or choose to share-crop with a landowner. Your decision could impact your overall profitability, as you'll need to generate sufficient revenue to cover the cost of expenses associated with paying rent or keeping up with mortgage payments. If you opt to begin your enterprise with share-cropping, you'll need to ensure you generate sufficient profits to provide the landowner with a large enough share to warrant your use of the land.

Know exactly what machinery and equipment you can use to get started and what technology could help improve your productivity and profitability overall. Once your final products are ready for sale, you should already have your packaging and shipping options prepared in advance.

Planning and Management

Willpower and hard work on their own won't make you a successful agribusiness owner. On the surface, farming seems to be mostly reliant on taking care of your crops or animals. But the reality is that good planning and ongoing management are the keys to any successful agriculture business.

Wherever possible, nothing should be left to chance. Every aspect of your current and upcoming season's revenue and expenditure should be already planned in advance.

If you're growing crops, create a cropping calendar and print it out. Keep your cropping calendar in clear view where you can always see at a glance what crops are due to be sown, when they're due to be fertilized and harvested, and what secondary or companion crops are due to be sown throughout the season.

Prepare plans for what you'll do to protect your income-producing crops in the event of inclement rain or heavy snow or unseasonal heat. Know how you'll mitigate risks of bug or insect infestations and plan for those unforeseen maintenance expenses that seem to always happen when you least expect them.

Know what crops also need to be planted as companion crops to repel bugs and pests or secondary crops that might also yield additional profits.

Seek out viable markets and potential customers to sell your crops to before they're due to be harvested. Develop your markets in advance and let them know what yields you anticipate and what volumes you hope to produce throughout harvesting seasons.

A critical part of your farm management process should also include bookkeeping and accounting. Sitting down to deal with the necessary paperwork associated with running a business makes many people cringe. Yet knowing exactly what revenue is coming into the business and what expenses are due is the key to staying on top of your business operations.

The vast majority of machinery and equipment you use on your farm will be tax deductible. All the tools and hardware you use to run your farm are also deductible. It's important you speak with a good accountant to be sure you're claiming the correct deductions as they pertain to your business.

If you're raising animals, the same planning and management rules apply.

Planning ahead for the ideal outcome for your agribusiness is a great way to keep on top of what needs to be done. But effective management also means putting contingency plans in place for when things might go wrong.

Insurance

A critical element in protecting your business interests against unforeseen circumstances is business insurance. The primary purpose of business insurance is to help control or mitigate unnecessary risks.

The types of insurance you should consider include:

- Property insurance;
- Farm motor vehicle and machinery insurance;
- Liability insurance, and;
- Life and disability insurance

Before choosing an insurance policy, take the time to read through your policy coverage options and know exactly what's insured. Some insurance companies may offer package policies that provide cover for a number of different aspects of your business operations under one policy.

Multiple Markets from One Product

Creative farmers have opportunities to spread their produce across multiple different markets. For example, let's say you grow cherries on your farm. Chances are you'll pick the ripe fruit once a year and sell it to grocery stores, local restaurants or eateries, or at your local farmer's market as fresh fruit sold by the pound.

However, once the season has ended and there are no more cherries to pick, your market opportunities end.

What if you only sold a portion of your crop as fresh fruit, but retained the rest to create into different products? Create cherry jam or jelly or preserves and store the finished product in air-tight jars that can be sold to local stores or directly to customers at farmer's markets at a higher rate than you could make selling the crop as fresh fruit.

The secondary products have a longer shelf life, so offer the opportunity to continue earning revenue from your crops throughout the year. There's also the opportunity to diversify your market base and expand your product's audience reach.

The same opportunity exists for almost every crop or animal you cultivate on your farm.

For example, if you raise goats the animals can be sold to meet the increasing demand for goat meat. Goats can also be a reliable source of goat's milk, which is also increasing in demand for its use in gourmet cheese-making.

As another example, you might raise quails that can be sold to meet the growing demand for quail meat. Quail eggs can also be sold at a secondary product, creating an additional stream of income from the same primary product.

Multiple Points of Sale

In layman's terms, generating multiple revenue streams is also known as not putting all your eggs in one basket. In business terms, it's known as broadening your debtor base.

Successful farmers and agribusiness owners know the importance of generating multiple streams of revenue from the same land, crops, or animals.

Having several different customers or market outlets or alternative points of sale to sell your completed produce or grown livestock or poultry ensures you always have income coming in from one source or another throughout the year.

A separate facet from generating more than one stream of income from the crops or animals you already farm is finding different points of sale. It's common for many market gardeners to remain loyal to one or two primary customers. Those large customers provide the vast majority of their revenue each year.

However, if your largest customer stopped ordering without warning, you'll suffer from the immediate loss of income unless you can find another large customer in a hurry.

Having a number of different options for selling the products you farm helps protect your income and overall profit. It also reduces your reliance on just one source of revenue, so you know your business is protected if someone stops ordering from you at any time.

Some examples for selling your products include:

- Contact local restaurants or grocery stores and take orders prior to harvest

- Set up a stall at a local farmer's market to sell directly to local customers
- Create a strong online presence and let loyal customers know that your products are available from your street-front presence or roadside stall.

Be creative about how you get your products out to customers. The objective is to ensure there is always demand for your goods from someone.

DIVERSIFY AND EARN HUGE INCOME

Farm diversification is commonly referred to as 'mixed farming' in agricultural communities. Diversifying any farm's resources or commodities can often provide more income stability throughout the year.

No matter what sized farming operation you operate, the opportunities available to diversity your farm's income are limited only by your location, the size of your available land or buildings, the climate, and your imagination.

Profiting from Under-Utilized Areas

Not every farmer utilizes every available square foot of arable farming land available to them to produce the goods they require to generate an income. If you're still new to farming and establishing your land, it's often necessary to begin small and expand as profits rise.

In the event that you have under-utilized land or buildings available on your land, the opportunity exists to use those areas to help boost your overall revenue.

Sub-leasing or Share Farming

In some cases, it may be possible to sub-lease land or unused building to another farmer for their own use. Some farmers will happily lease a portion of your land if they are experimenting with a small crop to test its viability. Others will be willing to lease land off you in an effort to increase their own crop yields.

If you choose to lease a portion of your land, be certain to prepare a legal agreement between you and the other farmer that clearly defines all payments, terms and conditions of the agreement.

Agistment

An agistment agreement may also be a useful way to generate additional income from unused areas of land.

Neighboring farmers may be willing to lease your unused pasture lands or fields to agist their livestock. Recreational horse owners also need available agistment land to keep their horses.

Your agistment agreement needs to be clearly worded to define payment terms and conditions. It also needs to verify what's expected of you, as some agistment agreements may include feeding and providing adequate shelter from weather conditions for animals being agisted.

Profiting from Small-Scale Dual Farming

The primary producing areas of your farm may be dedicated to raising animals or growing crops, but there is always the possibility of putting other areas of your property to good use with profitable small-scale dual farming options.

Many types of crops grow remarkably well as dual-purpose crops that have the potential to increase overall net crop returns. Some farmers may adopt the practice of growing two or more crops across the same piece of land during the same growing season.

Alternatively, double-cropping is the practice of planting a second crop in the same piece of land after the first crop has been harvested, or sometimes sown amid the first crop before it has been harvested. The result allows the farmer to take advantage of two separate harvesting seasons, which can improve revenues throughout the year.

Other crops may benefit from companion plantings, which also offer the opportunity to increase net crop returns.

Aside from growing multiple crops across the same farm, the opportunity also exists to raise smaller animals that don't require huge tracts of land to raise.

As an example, farming quails can offer a number of opportunities for many farmers. The increase in demand for quail meat from restaurants across the country can provide marketing opportunities. Selling quail eggs can provide a secondary source of income from the same endeavor.

Quail meat and quail eggs are an essential part of Mexican cuisine, but they're also commonly used in Vietnamese and Chinese cuisine. They're also increasing in popularity across fine dining restaurants.

Quail birds mature significantly faster than other types of poultry, reaching maturity in as little as six weeks instead of up to six months as with chickens. They also require much less space to raise as compared to turkeys or chickens.

It's common for many farms to maintain dams to provide water for livestock. In Australia, savvy farmers raising sheep or cattle as a primary farming product have also been using their dams to farm freshwater crayfish, or 'yabbies' as they're called Down Under to sell as a secondary stream of income for many years.

The increase in demand for crayfish across American restaurants provides additional marketing opportunities for some farmers living in areas where dams can be easily stocked with freshwater crawfish. Before stocking a dam with crayfish, take the time to check your state's aquaculture regulations for farming.

Profiting from Companion Planting

No matter what your primary profit crop might be, there are always opportunities for dual-crop farming. A hundred years ago farmers and home gardeners understood the importance of companion planting to repel plants and promote healthy growth of primary crops.

These days, farmers and market gardeners have come to rely on chemical pesticides and herbicides, as well as chemically-based fertilizing agents to keep crops safe.

Yet companion planting offers any agribusiness owner the opportunity to take advantage of dual-crop farming. Your primary crop is protected from pests and diseases by the presence of a suitable companion plant. Your farm revenue is then boosted by harvesting the companion crop as a secondary source of additional income.

The challenge is determining the correct companion plants to sow in order to optimize growth of your primary crop and add extra income to your business's bottom line. Here are some examples of profitable companion plants that could help improve your business profits.

Lavender

Lavender is such a versatile crop that it's possible to produce above-average profits per square foot of growing area. Many farmers plant lavender to help repel moths, slugs and deer. Cruciferous vegetables, such as broccoli or cabbage benefit from having a lavender plant nearby. Fruit trees that may be susceptible to attack from moths can also benefit by a nearby lavender bush.

Fresh lavender flowers can be sold as decorative arrangements or for future use in producing lavender oil. Dried lavender flowers are also popular with crafters and florists for use in a number of decorative arrangements. Dried lavender is also a popular ingredient in a number of beauty, skin care and aromatherapy products, including shampoo, soap, candles, sachets, and other herbal arrangements.

Lavender is easily grown in garden beds away from other income-producing crops, as it has a lovely decorative effect. It can also be grown as a border plant around other crops to provide protection.

Garlic

Garlic is the ideal companion crop for many market gardeners growing other types of vegetables. Garlic is known to be a natural, organic deterrent for many pests and fungus types, so it's a good crop to scatter through any garden.

Some common pests that are deterred by garlic include aphids, ants, snails, cabbage loopers, spider mites, codling moths and fungus gnats. Garlic has also been known to repel rabbits and even deer.

Garlic plants take very little space within a garden bed it also grows in most conditions as long as it has plenty of sun. Growing garlic is also a very low maintenance way to improve the flavor of many types of vegetables.

There are a number of crops that grow exceedingly well with garlic sown alongside them. These include fruit trees, dill, beets, kale, spinach, carrots, potatoes, tomatoes, peppers, broccoli, cauliflower, cabbage, and eggplant.

Gourmet Mushrooms

Mushrooms make the ideal specialty crop for many farmers, as they're grown indoors and are capable of producing an extremely high return per square foot used. If you have unused space in a barn or other indoor space, growing mushrooms could help improve your agribusiness yield and overall profit during seasons when other crops aren't producing as heavily.

Specialty mushrooms, such as shiitake and oyster mushrooms are perhaps the most profitable per square foot of growing area. Both types of specialty mushrooms can be sold dried, but the majority can be sold fresh, which is good news for local growers supplying grocery stores, restaurants, or individual customers direct at farmer's markets.

Keep in mind that some plants may suffer if planted anywhere near garlic. These include peas, beans, asparagus, parsley and sage.

Chamomile

Chamomile is an excellent companion plant for a number of reasons. Steep the flowers in water to create chamomile tea and spray the mixture on seedlings to prevent a fungal infection known as 'damping off'. Chamomile is also known to have antibacterial and anti-fungal properties that help protect plants prone to fungus, mildew or mold problems.

Many farmers use chamomile as companion plants to help improve the growth of a number of vegetables, including onions, cucumbers, kale, Brussels sprouts, cauliflower, beans, and cabbage. Chamomile also improves the flavor of mint and basil, so if you're growing herbs chamomile can be used as a companion plant there as well.

Profiting from Herbs

The demand for locally grown organic herbs has increased dramatically in the last decade. More people are using fresh and dried herbs in home cooking, resulting in an increase in availability of a broad range of herbs in grocery stores and local farmer's markets.

Herbs are also increasingly used in beauty and skin care products, as well as within handcrafted candles, aromatherapy products, essential oils, and soaps. Restaurants always need reliable suppliers of organic herbs.

Medicinal herbs are being used as natural alternatives to help improve and maintain health. Demand for high quality herbs from direct consumers and from manufacturers within the herbal product industry has meant that the number of commercial medicinal herbal farms available struggle to meet the ever-increasing need.

The beauty of adding herbs into your crop production schedule is that they take relatively little space to cultivate. Specialty herbs can also yield higher profits per square foot of growing space than many other types of vegetables.

Profiting from By-Products

No matter what your primary source of agribusiness income might be, there is always an opportunity to generate additional profits from by-products you already generate.

Creating a secondary income stream from products you already use or generate around your farm can boost your overall profits and help reduce your overheads at the same time. The key is finding the right by-products to suit your needs.

Compost

Any good home gardener knows the importance of adding compost into the soil before planting a garden bed. The same is true for any good market gardener. Yet both home and market gardeners often head into their local garden supply store or nursery to buy commercially-manufactured compost. You already have all the ingredients you need right on your farm to generate high-quality, organic compost that every gardener needs.

Food scraps, lawn clippings, dried leaves, saw dust, paper towels, shredded newspaper, drier lint, coffee grounds, egg shells, and plenty of other organic household scraps can be ideal to add as ingredients into any healthy compost mixture.

Besides, using your own organic compost also reduces your operating costs, as you won't need to buy any to enrich your soil each year. If you make more compost than your garden needs, you can bag it up and sell it to small garden centers.

Fertilizer

If you raise animals on your farm, you have the opportunity to cash in on their by-products. Manure can be a great way to fertilize any garden. Bag and sell poultry or livestock manure to the public or to small garden centers to help supplement your farming income.

Bees

Without bees to pollinate their crops, many farmers would go broke very quickly. While some farmers plant flowering shrubs or trees on their land to encourage bees to create hives on their properties, others are forced to hire apiarists to bring in bees on demand to fertilize their crops.

Keeping bee hives on your property offers multiple income-producing opportunities: you save the cost of hiring an apiarist to pollinate your own crops, you can hire out your bees to nearby farmers on-demand, and you can sell the honey they create.

Worms

Let's face it, any healthy garden needs worms to help aerate the soil around crop roots. Farming worms can be done on a small scale as a secondary source of income for your agribusiness. Garden stores may consider buying your worms for their gardening customers. Fishing stores also need a steady supply of worms to sell to fishermen to use as bait.

BUILDING A MILLION DOLLAR AGRIBUSINESS

Before we look more closely at the possibility of building a million dollar agribusiness, it's important to consider economy of scale.

It's possible to purchase an established farm or ranch on 1,000 acres of prime land, complete with farming machinery and tools and spend more than a million just getting your agricultural business started. By the time you spend money on cattle or livestock or seeds to get your crops sown, you've spent even more. In this event, your business has already cost you seven figures just to get going.

However, for the purpose of this guide we'll focus primarily on agricultural businesses that began as small-scale operations and subsequently built up to become successful agribusinesses generating millions of dollars in revenue each year.

CASE STUDY #1: Environmentally-Friendly Permaculture Farming Success

Joel Salatin has been hailed by Times magazine as "the world's most innovative farmer" for his incredible success with environmentally-friendly farming practices.

His family purchased some of the most eroded, abused farm land available in the Shenandoah Valley in Virginia and set about turning it into a profitable farming operation. The family planted trees, dug ponds, and built huge compost piles to replenish and revitalize worn-out, badly degraded soil.

Using a permaculture approach to promote sustainable living and land use, the family designed their farm to mimic patterns and relationships found in nature. Over time, the 220-hectare property began to recover and became fertile grazing land with arable farming soil.

Each day, the family moves cows in a rotational grazing system and invented portable sheltering systems that allow them to produce animals on perennial prairie polycultures.

Today Polyface Farms sustains a thriving pig, poultry, cattle, rabbit and forestry farm that boasts an annual turnover of around $2 million. The Polyface Farm supplies fresh, locally grown produce to around 50 restaurants, 5,000 families, and 10 retail outlets, along with selling direct to the public from their on-farm shopfront.

CASE STUDY #2: Cherry Grove Organic Farm

Matt Conver was not from a family of farmers, but he had a vision. He began his agricultural business on 19 acres of land three miles south of Princeton, NJ. His intention right from the beginning was to grow a diverse selection of fresh, organically grown produce without the need for toxic, synthetic chemicals.

Conver is a part of a growing trend of people who prefer to lease their farmland from property owners rather than repay a hefty mortgage. He promotes healthy soil by planting soil-improving crops and with use of balanced farming practices to ensure his produce is always the best it can possibly be.

The farm's produce is sold direct to consumers at two farmer's markets in New

Jersey, as well as generating revenue through Community Supported Agriculture (CSA), through which he feeds around 300 families each year.

Demand for the farm's organic produce has grown to the point that Conver's dad, mom, sister and nephew also help out at farmer's markets and around the farm. The farm has become so profitable that Matt's wife, who is a trained social worker, is able to stay home with the kids.

CASE STUDY #3: Dairy Farming and Artisanal Cheese Making

Diana Giacomini Hagan grew up on her parents' dairy farm in California, so she was already familiar with the lifestyle. Yet she decided to leave the farm and go to college, pursuing a career in corporate America.

Before long, her father raised the idea of taking the milk produced on the farm to the public in the form of cheese. Yet he had no idea how to get it started.

Diana's sisters accepted the challenge, hiring a cheesemaker to help them produce their first classic-style blue cheese. After some years of trial and error, the family began to increase production to meet the growing demand for their products.

Before long, the cheese created on the farm became so popular that Diana left her corporate job in real estate to go back to the family farm as the farm's chief financial officer.

From humble beginnings as a dairy farming operation operated by Diana's mother and father, the farm now employs Diana and her two sisters, along with around 50 employees.

The business also manages 750 acres of pasture, milks 350 cows daily and produces revenues of more than $5 million from sales of three different types of cheeses produced on the premises each year. Diana estimates the farm's production levels are on track to see an increase of around 20% in the amount of cheese sold in the following year.

CASE STUDY #4: Crawfish and Rice Farming Success

Most people associate agricultural farming businesses as being limited to mid-west locations. Yet agribusiness in southern parts of the country is flourishing. Brothers Mike and Mark Fruge, the owners of Cajun Crawfish farms began as a small crawfish farming operation in the early 1980s. Early sales were made from the back of Mike's truck, selling their crawfish door-to-door in local restaurants. As sales began to soar, the brothers increased the size of the farm by adding more acres in an effort to produce more crawfish.

The brothers soon diversified their crawfish farming operations to include farming rice on a rotating basis. While crawfish farming began as the farm's primary income, the owners soon learned that commercial rice sales were an intensely lucrative way to continue earning money from the same crawfish ponds throughout the summer months.

The farm focuses on growing a healthy rice crop from March through to July. The field is seeded with crawfish in June. By the time the rice harvest is complete in August, the rice field is then drained and re-flooded again to become a crawfish pond for the rest of the year.

The crawfish are harvested from November through to July, creating a profitable secondary harvest through the winter months.

Today, the farm spans 1,500 acres or crawfish ponds and rice fields and has grown into a full-service seafood company covering most of the major markets across Louisiana, Texas, and Oklahoma.

CONCLUSION

At a time when so many people are shying away from agriculture due to financial concerns, there are plenty of others willing to get out on the land and prove that agriculture is the key to their success.

Besides, when it's operated well, an agricultural business is the ideal recession-proof business. Even when people are struggling through tough economic times, they still want healthy food.

Agribusinesses don't have to be large-scale operations in order to be profitable. There are also plenty of ways for any keen agribusiness owner to reduce operational costs by using natural ways to fertilize and replenish soil, control pests and weeds, and start seeds for the next season's crops.

The key to any agricultural business's success is innovation and effective management. America is home to many, many successful agribusinesses that provide food and jobs for a growing number of people right across the country.

If you're ready to leave the constant grind and stress of the corporate rat-race and head out to greener pastures to enjoy a more relaxed lifestyle, agriculture provides the ideal option.

Take some time to plan your business and create a strong structural foundation for your business model. Work on ways to make the best use of your land so you're able to diversify income opportunities and broaden your market.

You'll soon find you have the keys to developing a profitable, sustainable business that not only generates wealth, but also provides a satisfying lifestyle for years to come.

Other Available Books:

- In The Pursuit of Wisdom: The Principal Thing

- **Investing in Gold and Silver Bullion - The Ultimate Safe Haven Investments**

- Nigerian Stock Market Investment: 2 Books with Bonus Content

- **The Dividend Millionaire: Investing for Income and Winning in the Stock Market**

- Economic Crisis: Surviving Global Currency Collapse - Safeguard Your Financial Future with Silver and Gold

- **Passionate about Stock Investing: The Quick Guide to Investing in the Stock Market**

- Guide to Investing in the Nigerian Stock Market

- Building Wealth with Dividend Stocks in the Nigerian Stock Market (Dividends - Stocks Secret Weapon)

- **Bitcoin and Digital Currency for Beginners: The Basic Little Guide**

- Child Millionaire: Stock Market Investing for Beginners - How to Build Wealth the Smart Way for Your Child

- **Christian Living: 2 Books with Bonus Content**

- Beginners Quick Guide to Passive Income: Learn Proven Ways to Earn Extra Income in the Cyber World

- **Taming the Tongue: The Power of Spoken Words**

- The Power of Positive Affirmations: Each Day a New Beginning

- The Real Estate Millionaire: Beginners Quick Start Guide to Investing In Properties and Learn How to Achieve Financial Freedom

- **Business: How to Quickly Make Real Money - Effective Methods to Make More Money: Easy and Proven Business Strategies for Beginners to Earn Even More Money in Your Spare Time**

- Money: Think Outside the Cube: 2-Book Money Making Boxed Set Bundle Strategies

- **Marketing: The Beginners Guide to Making Money Online with Social Media for Small Businesses**

- How to Effectively Lead and Win: The Proven Leadership Strategies and Techniques

- The Quick Guide to Robotics and Artificial Intelligence: Surviving the Automation Revolution for Beginners

If you would like to share this book with another person, please purchase an additional copy for each recipient. Thank you for your support and thanks for reading this book.

www.ingramcontent.com/pod-product-compliance
Lightning Source LLC
Chambersburg PA
CBHW031546210526
45464CB00003B/1170